AIR SCIENCE

Active Science with Air

Science Action Labs

Written by Edward Shevick
Illustrated by Marguerite Jones

Teaching & Learning Company

1204 Buchanan St., P.O. Box 10
Carthage, IL 62321-0010

This book belongs to

_____

Cover art by Marguerite Jones

Copyright © 1998, Teaching & Learning Company

ISBN No. 1-57310-142-7

Printing No. 987654

Teaching & Learning Company
1204 Buchanan St., P.O. Box 10
Carthage, IL 62321-0010

TLC10142 Copyright © Teaching & Learning Company, Carthage, IL 62321-0010

Table of Contents
Science Action Labs

Dear Teacher or Parent,

The spirit of Sir Isaac Newton will be with you and your students in this book. Newton loved science, math and experimenting. He explained the laws of gravity. He demonstrated the nature of light. He discovered how planets stay in orbit around our sun.

Air Science can help your students in many ways. Choose some activities to spice up your class demonstrations. Some sections can be converted to hands-on lab activities for the entire class. Some can be developed into student projects or reports. Every class has a few students with a special zest for science. Encourage them to pursue some **air science** experiments on their own.

Enjoy these science activities as much as Newton would have. They are designed to make your students **think**. Thinking and solving problems are what science is all about. Each section encourages thought. Students are often asked to come up with their best and most reasonable guess as to what will happen. Scientists call this type of guess a **hypothesis**. They are told how to assemble the materials necessary to actually try out each activity. Scientists call this **experimenting**.

Don't expect the experiments to always prove the hypothesis right. These air science activities have been picked to challenge students' thinking abilities.

All the activities in **Air Science** are based upon science principles. Many are explained by Newton's laws. That is why Sir Isaac Newton has been used as a guide through the pages of this book. Newton will help your students think about, build and experiment with these activities. Newton will be with them in every activity to advise, encourage and praise their efforts.

The answers you will need are on page 64. You will also find some science facts that will help your students understand what happened.

Here are some suggestions to help your students:

1. **Observe carefully.**
2. **Follow directions.**
3. **Measure carefully.**
4. **Hypothesize intelligently.**
5. **Experiment safely.**
6. **Keep experimenting until you succeed.**

Sincerely,

Ed

Edward Shevick

What Is in Our Air?

Is Air Real?

You can't see air. You can't smell air. You can't even taste air. You can feel it while riding your bike or when you place your hand outside a moving car.

You can live a month without food. You can live a week or more without water. Without air, you couldn't last five minutes.

The air we must have is in an ocean extending up to 500 miles (805 km) above the Earth. Half of the air in our world is found in the first three miles above Earth.

Feeling Air

Here is a way to "feel" and "see" air.

1. Obtain a small aquarium or large jar and a glass.

2. Fill the aquarium or jar so that it is $2/3$ filled with water.

3. Lower the glass *mouth down* into the water until the bottom of the glass is at water level. You are actually compressing the invisible air. Did you feel the glass of air resisting while being lowered?

4. Now tilt the glass of air so that more water enters. Describe what you see.

_____

_____

Newton Challenge: Newton can pour one glass of air into a second glass of water while **both** are submerged. Can you duplicate Newton's pouring trick?

Name _____

Our Air Is Mainly Oxygen and Nitrogen

Air is about $\frac{1}{5}$ oxygen and $\frac{4}{5}$ nitrogen. The other gases in the air will be covered in the lab on page 8.

Oxygen is the part of the air we need. We burn it in the cells of our body to give us energy and warmth. Oxygen burns with gasoline to power our cars. Plants and animals could not live without oxygen.

You can prove for yourself that air is roughly $\frac{1}{5}$ oxygen.

1. Obtain a large bowl and fill it almost to the top with water.

2. Obtain a tall graduate or a tall thin jar.

3. Measure its total height. _____ inches (_____ cm).

4. Mount a candle on a cork. The candle and cork must fit inside the graduate.

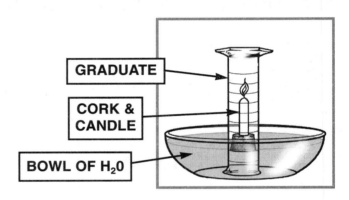

GRADUATE

CORK & CANDLE

BOWL OF H_2O

5. **Caution! Have an adult light the candle and place it and the cork on the water.**

6. Place the graduate or tall jar over the burning candle and cork. Push the graduate straight down into the water until the top touches the bowl. Describe what happens as the candle goes out.

_____

_____

_____

7. Measure the height of water in the graduate. _____ inches (_____ cm)

Example: Let's suppose the graduate was 15" (38 cm) high. Let's suppose the water rose 3" (8 cm). Now do the math for your graduate and water rise. Round off to whole numbers.

$$\frac{3}{15} = \frac{1}{5}$$ equals the amount of the graduate filled with water.

Your answer = _____

6

Name _____

Newton Explains: The empty graduate contained mostly nitrogen and oxygen. Nitrogen does not burn. Only the oxygen burned up. Water rose up the graduate to replace the burned oxygen.

After the burning, what was the main gas left in the graduate? _____

Per your math, what fraction of the graduate was filled with water? _____

What gas did this water displace? _____

Newton Apology: This experiment is not perfect. All the oxygen is rarely burned. Some burning gas is trapped inside the graduate. Some of the heated air escapes the graduate as it expands.

Newton Wants You to Research Nitrogen

Nitrogen is the key chemical needed to make the essential proteins of life. Nitrogen in the air dilutes the oxygen so things don't burn too rapidly. A candle in pure oxygen would burn in $1/10$ the normal time. Nature uses both lightning and bacteria to convert nitrogen in the air into plant fertilizers.

You can learn more about nitrogen through research. Here are some possible topics suggested by Sir Isaac.

Why is nitrogen called an "inert" gas? _____

How do bacteria and certain algae "fix" air nitrogen into a useful powder? _____

_____

How does lightning "fix" air nitrogen into a solid? _____

_____

What is the role in the nitrogen cycle of "decomposing" bacteria? _____

_____

How do volcanoes affect the Earth's nitrogen? _____

_____

Can you make a chart of the complete nitrogen cycle? _____

Name _____

A Gas Guessing Game

Newton wants you to know about the other gases in the air. Since most of the air is **nitrogen** and **oxygen**, the other gases are found in very, very small quantities. Scientists call these small amounts **trace** gases.

Here is a list of air gases. Can you identify them by Newton's clues? Fill in the left side of the chart.

AIR GAS CHART

	Gas Name	Chemical Symbol	Percent by Volume	Uses and Clues
1		N	78	Proteins and fertilizers; lightning affects it
2		O	20.95	All living things need it; fuel in rockets
3		AR	0.93	Used in electric light bulbs
4		CO_2	0.03	Found in soda pop; used by plants in photosynthesis
5		NE	Trace	Used in advertisings signs
6		HE	Trace	Very light gas used in balloons and blimps
7		KR	Trace	Superman doesn't like this
8		XE	Trace	Used in electronic flash bulbs
9		H	Trace	Part of water molecule
10		H_2O	0 to 4.0	Necessary for all life. Amount in the air can vary every day.

Air Matters

Newton Parachutes to Earth

Air is matter. It takes up space. It has weight. Air in the form of wind drives sailboats. Our friend Newton wouldn't dare make a parachute jump if air was not matter. Let's experiment with a plastic bag of air.

1. Obtain a sturdy plastic bag.

2. Tie it securely at the opening.

3. Now push on the "empty" bag. It pushes back because air is matter and takes up space.

I'm Falling for You

Falling objects, like parachutes, are affected by the resistance of our ocean of air. Let's use a paper plate to experiment with air resistance.

1. Obtain a 6" (15 cm) or larger paper or plastic plate.

2. Stand on a chair and prepare to drop your plate from positions A, B and C (pictures at the right).

   Hypotheses Time: Before dropping the plates, predict which ones will have the **least** flip-overs. A flip-over is a complete turn in the air.

   Your prediction: _____

3. Try each position a few times. Which position had the least flip-

   overs? _____ Try to explain why in terms of air resistance.

   _____

   _____

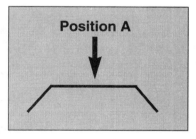

Position A

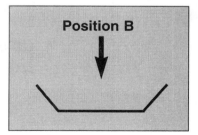

Position B

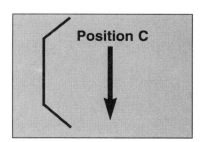

Position C

Name _____

The Stubborn Juice Can

NAIL HOLE IN BOTTOM OF CAN CLOSE TO RIM

1. Obtain a full, individual-size juice can; a hammer and a large, clean nail.

2. Punch a hole **near the rim** on the **bottom** side of the can.

3. Try pouring the juice out of your nail hole. Describe your results. _____

_____

4. Punch a second hole on the **opposite** side of the can.

5. Try pouring the juice out again. Describe your results. _____

_____

Newton Air Wisdom: You must allow air into the can to force the juice out.

Trapped Water

1. Obtain a small jar with a lid. Baby food jars work fine.

2. **Remove the lid.** Punch five or six holes completely through the lid.

3. Fill the jar with water and place the lid on **tight**.

4. **Go outside**, or stand over a sink, and turn the jar over completely as shown. Describe what happened. _____

_____

Explain why the water stayed in the jar in terms of air pressure. _____

_____

5. Now **tilt** the jar so that it is no longer vertical. Describe what happened.

_____

Explain what happened in terms of air pressure. _____

_____

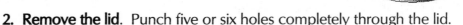

10

Balloons: Elastic Bags of Air

Newton Still Plays with Balloons

The dictionary defines a *balloon* as "an inflatable rubber bag". It keeps air in and is often used as a brightly colored child's toy. Giant balloons, filled with hot air, can let you travel around the world.

Newton uses balloons to demonstrate science principles. Your lungs are basically two balloons filled with air. As air escapes from a full balloon, it demonstrates Newton's law of action and reaction.

HOT AIR BALLOON USED IN 1783

The Impossible Balloon

1. Push a deflated balloon into a soda pop bottle.

2. Stretch the balloon mouth over the mouth of the bottle as shown.

3. Try to blow up the balloon. Describe what happened. _____

_____

_____

Try to explain why you failed to blow up the balloon.

What is happening to the air trapped in the bottle? _____

_____

_____

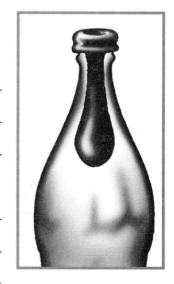

Name _____

Building an Air Puck

Balloons can be used to store energy. An air puck is a device that uses stored air to cut down on friction.

Here's how to build an air puck.

MODEL AIR PUCK

1. Obtain a balloon, a **small** spool and a **thin** piece of **smooth** wood about 2$\frac{1}{2}$" (6.33 cm) square.

2. Drill a hole through the exact center of the wood square.

3. Glue the spool to the wood so that the hole in the spool is directly over the hole in the wood square.

4. Blow up your balloon to its maximum size and attach it to the spool.

5. Place your air puck on a **smooth** surface and give it a slight sideways push.

Newton's Air Puck Challenge

There is more than one way to build an air puck. You could use plastic or thick cardboard as a base. You could have a round or triangular shape. You could use a one-hole stopper instead of a spool.

Be different! Challenge your classmates!

Name _____

How Much Does the Air in a Balloon Weigh?

You could weigh a quart of milk indirectly by weighing yourself, drinking the quart of milk and weighing yourself again. The difference would be the weight of the quart of milk.

Weighing the air in a balloon is difficult but not impossible. Here is how it can be done.

1. Obtain a balance, a ruler, a large paper clip and a balloon.

2. Weigh the **empty** balloon and the paper clip together.

   Empty balloon and paper clip weigh _____ ounces (grams).

3. Blow up the balloon to about 10" (25 cm) in diameter.

4. Twist the full balloon's neck and seal it with the paper clip so that no air escapes.

5. Again weigh the **full** balloon and attached paper clip.

   Full balloon and paper clip weigh _____ ounces (grams).

Find the weight of the air in the balloon by subtracting the empty weight from the full weight.

Weight of air in balloon equals _____ ounces (grams).

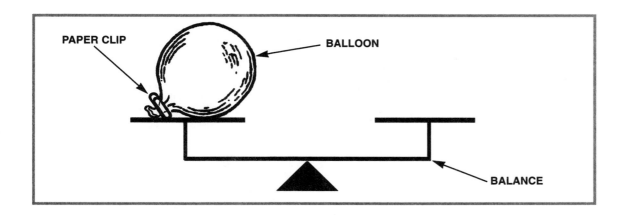

Name _____

Balloon Puzzler

Newton Explains the Puzzle

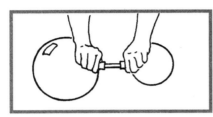

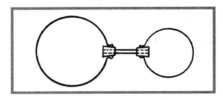

Two balloons are connected to each other by one-hole stoppers and glass tubing. The smaller balloon is about the size of a grapefruit. The larger balloon is the size of your head. The necks of the balloons are squeezed so that air cannot flow from one to the other.

What direction would the air flow within the balloons? Assume the balloon necks are released and the air is free to move as it pleases.

What do you predict will happen?

_____

_____

How to Build Newton's Puzzler

1. Obtain two **new** balloons that can be blown up at least 12" (30 cm).

2. Obtain a 4" (10 cm) section of glass tubing and two one-hole stoppers.

3. Place the balloons around the **larger** end of the stoppers as shown.

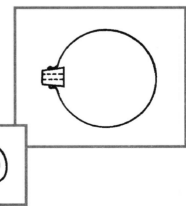

14

Name _____

4. Place the glass tubing in the **smaller** end of one of the stoppers. Use soap or water to lubricate the glass so that it is easy to insert. The glass need only go about 1/4 of the way through the stopper.

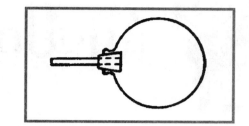

5. Blow through the glass tubing until the balloon is about the size of a grapefruit.

6. Squeeze the neck of the balloon so the air cannot escape.

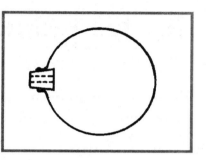

7. Have a friend blow directly through the other stopper to fill the second balloon. It should be about the size of your head. Have the friend squeeze the balloon neck so the air cannot escape.

8. Attach the two balloons together by inserting the free end of the glass tubing into the stopper attached to the larger balloon. **Keep squeezing both necks** so no air can flow from one to the other.

9. Release both balloons to let the air flow freely between them.

Describe what happened. _____

_____

_____

Newton Gets the Last Word

Congratulations! You've built and tested another Newton puzzler. Maybe it didn't turn out the way you predicted it would. Don't worry about being wrong. Scientists are often wrong, but they learn from their mistakes.

Name _____

Air Under Pressure

Newton Lectures on Air Pressure

BODY AIR PRESSURE INSIDE and OUT

15 POUNDS ON EVERY SQUARE INCH

You live at the bottom of an ocean of air. This air provides the oxygen you need to survive.

The layer of air around the Earth is called the atmosphere. It mainly exists in the first few miles above the Earth's surface.

The pressure at the Earth's sea level surface is around 15 pounds (6.75 kg) on every square inch (centimeter). Think of how many square inches (centimeters) there are on the surface of your body. The air pressure would crush you if there was no air inside of you pushing out.

In all the experiments below, the explanation is the pressure that the weight of air exerts.

AIR PRESSURE FACTS

- The air constrained in an average drinking glass has the same weight as an aspirin tablet.
- Total air pressure on a human body can be over 10 tons (9 tonnes).
- Air pressure per square inch (centimeter) is equal to the weight of a bowling ball.

Air Pressure Is Exerted in All Directions

Living under an ocean of air is similar to living under an ocean of water. The deeper you swim in an ocean, the more pressure is on your body. You mostly feel this pressure in your ears.

16

Name _____

The pressure of water acts in **all** directions. Turning your head in any direction does not help the pain in your ears.

Air pressure is also exerted in **all** directions. Demonstrate this for yourself with a card and glass of water.

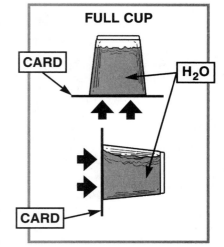

FULL CUP

CARD — H_2O

CARD

1. Fill a glass with water.

2. Place a playing or index card over the glass top.

3. Do the next step over a sink, a bucket or outside.

4. Turn the glass and card slowly upside down as shown.

   Describe your results._____

   _____

5. Now slowly turn the glass sideways as shown above.

   Describe your results._____

   _____

What do these experiments teach you about air pressure?_____

_____

Experimenting with Air Pressure

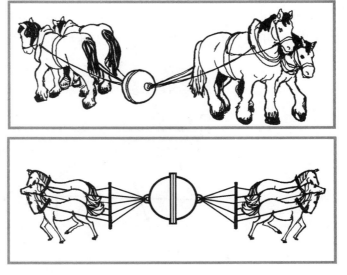

Way back in 1654, a scientist named Otto von Guericke demonstrated the power of air pressure. He fitted two copper hemispheres together and pumped the air out of the inside. He attached horses to each hemisphere. The horses couldn't pull the hemispheres apart due to air pressure.

Name _____

You probably don't have horses available. Try this version of Von Guericke's experiment.

1. Obtain two toilet plungers or two suction cups that are the same size.

2. Moisten the edges slightly and press them together **firmly**.

3. Now try to pull them apart.

Describe the results. _____

_____

Assume that you removed most of the air from within your plungers. What is holding the plungers together? _____

_____

Newton Loves to Break Pencils

Newton wants to show you how to break a pencil using air pressure.

1. Place a pencil on a table so that half of it sticks over the edge as shown.

2. Place a folded newspaper over the part of the pencil on the table.

3. Use a ruler to give the extended pencil a **sharp quick** blow.

The pencil breaks. It is not only the weight of the paper that causes this. It is the air pressure pushing down on the paper's surface. Inertia also helps.

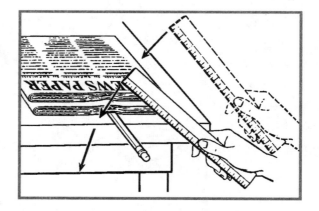

18

Name _____

Air Pressure Fun

Air Pressure Background

HOW A BAROMETER WORKS

VACUUM
34 FEET OF WATER OR 30 INCHES OF MER-CURY
POOL OF MERCURY OR WATER

Air has weight. The total weight of air at sea level is 14 pounds (6.3 kg) per square inch (centimeter). We usually round it off to 15 pounds (6.75 kg).

We can measure air pressure with a mercury barometer. Mercury barometers work because air pressure can support 30" (76 cm) of heavy mercury liquid. If you tried to make a water barometer, air pressure could hold up a 34' (10.2 m) column of water.

The Soda Pop Bottle Crush

SODA POP

1. Obtain a 2-liter plastic bottle.

2. **Carefully** pour very hot water into it until it is 1/2 full.

3. Swish the hot water around for about one minute.

4. Pour out the water and **quickly** and **firmly** screw on the cap.

Describe what happened to your plastic bottle. _____

_____

The hot air in the bottle cooled and took up less space. Can you explain what caused the jar to collapse? _____

_____

The Metal Can Crush

Now Newton wants to help you crush a metal can using air pressure.

Do this experiment only with the help of an adult!

Name _____

1. Obtain a pint or quart **metal** can that has a small screw cap. **Make sure the can is clean.**

2. Pour ¹/₂ (120 ml) cup of water into the can.

3. Place it on a stove until you see steam coming out.

4. **Turn off the stove. Carefully** place the lid on tightly.

5. Wait a few minutes and you will hear strange noises coming from the can.

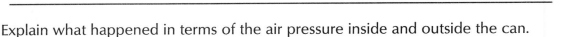

What happened to the can? _____

_____

Explain what happened in terms of the air pressure inside and outside the can.

_____

_____

Newton's Air Pressure Math

You have learned that air pressure is roughly 15 pounds (6.75 kg) per square inch (centimeter). Use that fact to compute the following pressures:

1. Total air pressure on all six sides of a 10" (25 cm) cube.

   _____ Total pounds (kg) of pressure

2. Total air pressure on the **top only** of a dollar bill.

   _____ Total pounds (kg) of pressure

3. Total air pressure on the floor of your classroom.

   _____ Total pounds (kg) of pressure

4. Total pressure on the outside surface of a standard basketball.

   _____ Total pounds (kg) of pressure

Newton Hint: You can mark off inches (centimeters) or use this formula for the area of a sphere. $A = \prod D^2$

Newton's Soda Pop Bottle Problem

Newton's Problem

Newton once worked in a plant that filled small-necked gallon bottles with pop. One day a worker forgot to add the flavoring and 500 gallon bottles of pop were ruined.

Someone had to empty all those gallon bottles into a sink one by one. Naturally, they called in Newton to discover the fastest way to empty them. Using Newton's method, they were able to empty each bottle in less than 15 seconds.

Newton Wants You to Hypothesize

Can you find a fast method of emptying a gallon (liter) bottle? Of course, you are not allowed to break the bottle. Before actually trying this puzzler, can you come up with some speedy ways to empty the bottle?

List your ideas.

1. _____

_____

_____

Name _____

2. _____

_____

_____

3. _____

_____

_____

4. _____

_____

_____

Experiment Time

Try to beat Newton's 15 seconds. Please read the **caution** section below.

Caution! Be careful doing this experiment so that you don't accidentally drop the gallon (liter) bottles and hurt yourself. Do the experiment outdoors or over a convenient sink.

There are many ways to approach a science problem. You can use the trial and error method and just try various emptying techniques. Or you can start first by thinking and trying to outsmart Newton. Good luck in beating Newton's 15 seconds.

NEWTON'S
ACTION LAB
Air
Science
11

Measuring Air Pressure: How Barometers Work

Newton Lectures on Air Pressure

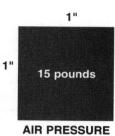

1"

1"

15 pounds

AIR PRESSURE AT SEA LEVEL

You live at the bottom of an ocean of air. At sea level, the air pressure is roughly 15 pounds (6.75 kg) per square inch (centimeter). This air pressure affects you in many ways. If you did not have air inside of you, you would be crushed.

Measuring air pressure is important for weather predictions. When air pressure falls rapidly, we are likely to have a storm. High or rising air pressure could mean fair weather ahead.

Measuring air pressure is done with a **barometer**. The word means "pressure-measure." The first barometer used a tube of mercury 30" (76 cm) high set in a bowl of mercury. Mercury is used because it is a very heavy liquid. If you used water instead of mercury, you would need a 34' (10.2 m) water tube.

Way back in 1646, a scientist named Otto von Guericke made a 34' (10.2 m) long brass tube filled with water. He placed a section of glass at the top and floated a doll on the water. The entire water barometer was attached to the side of his house. People would marvel when the doll moved up and down as air pressure changed.

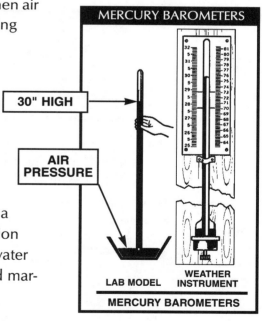

MERCURY BAROMETERS

30" HIGH

AIR PRESSURE

LAB MODEL

WEATHER INSTRUMENT

MERCURY BAROMETERS

Aneroid Barometers

Mercury barometers give very accurate air pressure measurements. However, they are expensive, heavy and awkward to move around.

Name _____

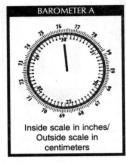

BAROMETER A

Inside scale in inches/
Outside scale in
centimeters

BAROMETER B

Inside scale in inches/
Outside scale in
centimeters

Most air pressure readings are taken with an **aneriod** barometer. Instead of mercury, aneroid barometers use a flexible metal box that has had some air removed. As air pressure changes, the flexible metal box shrinks or expands. A dial system records the air pressure on the marked scale. Aneroid barometers are simple to use, move about and read.

The normal barometer reading at sea level is 30" (76 cm). Observe barometer A. It shows an arrow pointing at the normal air pressures.

Observe barometer B. Record the air pressure it shows.

_____ inches (_____ cm)

Try to obtain an aneroid barometer. Take air pressure readings in inches (centimeters) for five days to observe daily changes.

Newton Hint: You will find a second moveable pointer on your aneroid barometer. By setting at the starting air pressure, you get a better idea of changes in pressure.

Day 1 _____   Day 2 _____   Day 3 _____   Day 4 _____   Day 5 _____

Newton's Cheap Barometer

Newton wants you to make a model barometer. It works but it has problems. It's not accurate. It's not reliable. It changes when moved. It is affected by heat as much as by air pressure. Build it for fun, but don't expect the Weather Bureau to buy it from you.

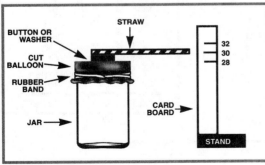

1. Obtain a tall jar or tennis ball container, a straw, a balloon, a button and a thick rubber band.

2. Assemble as shown. Use the rubber band to hold the cut balloon over the open end of the container or the can.

3. Glue a button or washer to the center of the balloon. It should stick out about $1/4$" (.6 cm) above the top of the container or the can.

4. Glue or tape a straw to the button or washer as shown.

5. Mount a piece of thick cardboard on any kind of stand so that your straw points to it.

6. Mark the "normal" straw position as 30. Mark a space above as 32 and an equal space below as 28. Test and record day-to-day barometric differences.

Day 1 _____   Day 2 _____   Day 3 _____

Winds: Air in Motion

Newton Explains the Wind

Wind is air in motion. Winds are in constant motion above land and oceans. Winds clear fog from our harbors and smog from our crowded cities. The wind can move sailboats and draw artistic patterns on sandy deserts. Winds chase clouds across our skies.

Air motion and energy can be traced to the sun. The air around the Earth's equator gets heated the most. Hot air expands, gets lighter and rises. Cooler air moves in from polar regions to take its place. These world-wide air movements are called **trade** winds. Trade winds are also affected by the Earth's rotation.

Many areas near oceans and lakes develop offshore-type winds. This is because land masses absorb and radiate sunshine energy better than water masses.

BIG WINDS

Hurricanes are tropical storms in which winds can reach speeds of 74 to 200 mph (119 to 322 km). The entire hurricane can stretch over 300 miles (483 km) and may move over the Earth's surface at an erratic 10 to 30 mph (16 to 48 km). Hurricanes form over water. Our east coast hurricanes form in the tropical Atlantic Ocean, the Caribbean Sea and the Gulf of Mexico.

Tornadoes are violently rotating air columns. They form the familiar funnel shape as they descend from a thunderstorm. The funnel of a tornado is only a few hundred yards (meters) wide and can reach speeds up to 300 mph (483 km). Tornadoes rotate counterclockwise in our hemisphere and are most frequent in spring.

Name _____

Measuring Wind Speed and Direction

ROOFTOP WEATHERVANE

Knowing the wind speed and direction is very important for weather prediction. Wind direction is determined by a **weather vane**. The simplest weather vane has a slim point at the front and a flared tail at the back. The tail is often arrow-shaped with two broad surfaces separated by about 20 degrees. This shape forces the weather vane to face into the wind. Thus, if the vane points to the northeast, the wind is coming from the northeast.

Newton never bothered with weather vanes. He just wet a finger and held it high. The cool side of the finger faced the wind.

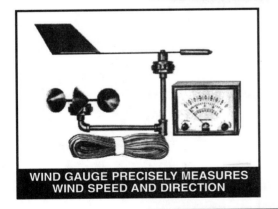

WIND GAUGE PRECISELY MEASURES WIND SPEED AND DIRECTION

Wind speed is measured by an **anemometer**. Most anemometers have three or four cone-shaped cups which rotate in the wind. The speed of rotation is proportional to the speed of the wind. The energy of the rotating cups is transmitted electrically to an indicator calibrated in miles per hour (kilometers) or any desired unit. This is done in much the same way that the moving wheels on an auto transmit a signal to an instrument that tells your car's speed.

Newton never bothered with anemometers. He just observed how flags flew, smoke drifted or tree branches bent.

Wind Fun

Cracker Windmill

Let a cracker be a windmill. You can be the wind. Obtain some square-shaped crackers. Hold the cracker **gently** between your thumb and middle finger as shown. Blow on one end of the cracker to spin it. You may need a little practice to perfect your windmill.

Name _____

Make a Pinwheel

1. Obtain a 6" (15 cm) square of heavy paper or thin cardboard. You will also need a short pin or nail, a pencil with an eraser and scissors.

2. Draw two lines joining the opposite corners.

3. Place an *X* on four corners as shown.

4. Cut along the lines to within an inch (centimeter) of the center.

5. Bend each corner containing an *X* (every other corner) so that all *X* corners lay over the center.

6. Push a pin or thin nail through the center of your pinwheel and into the eraser.

7. Try spinning your pinwheel with your breath, a fan, the wind or while running.

Steps 2-4

Step 5

Step 6

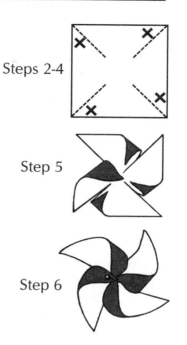

Newton Pinwheel Challenge: Try building a better pinwheel using different materials, patterns or construction.

Winds as an Energy Source

Up to 2% of the sunlight received by the Earth is converted to wind energy. Windmills to harness this energy can be traced back to ancient Persia. The Netherlands use many windmills today for grinding grain and pumping water. Windmills can also turn generators to produce electricity.

Electric energy from wind is simple to obtain and does not pollute the environment. There are some drawbacks to producing windmill electricity. It is often more costly than other energy sources. Since winds are not constant, a complicated and expensive storage system is needed during windless periods. Some environmentalists consider massive windmills ugly.

Name _____

Smog Lab: Counting Air Garbage

Newton's Smog Lecture

Take a deep breath. If you are a city dweller, the chances are very high that your lungs are filled with more than fresh, pure air. You have just taken in some air garbage ranging from dust to soot, from pollen to bacteria and from ozone to ammonia. You may have slowed up the **cilia** (hair-like structures) inside your lungs and clogged your **alveoli** (tiny grape-like air sacs). The pollutants you breathe could disintegrate nylons, corrode metals and kill plants. **Smog** affects your breathing, irritates your eyes and causes headaches.

WHAT'S IN THE AIR?	
Nitrogen	78%
Oxygen	21%
Rare Gasses (Argon, Neon, Helium)	1%
Carbon Dioxide	.03%
Water Vapor	0 to 4%

The average person needs to take in about 30 pounds (13.5 kg) of fresh air every day. Clean air is 78% nitrogen which goes in and out of our bodies unchanged. About 21% of the air is the oxygen we need to burn our food for energy and warmth.

We all live at the bottom of an ocean of air. Obtaining the right quantity of air is no problem. To extract the right quality of air from the chemical "soup" in the air is much more difficult. Very few places in the world are free of air pollution.

Help! I'm Trapped in Jelly

Most air pollutants are invisible. They are either in gas form or tiny particles that can't be seen without a microscope. These particles in the air are called **particulates**. They can be either solid or liquid materials.

Scientists measure particulates in a unit called **microns**. A micron is $1/1000$ of a millimeter or $1/25,000$ of an inch. You can't see viruses. They are less than one micron. You can see particles over 10 microns with your naked eye.

30

Name _____

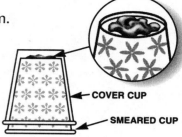

1. Obtain eight paper cups and either some petroleum jelly or cold cream.

2. Smear petroleum jelly or cold cream as evenly as you can on the outside bottom of four cups **only**.

3. Cover each of the four smeared cups as shown with the four remaining cups to protect them. See drawing.

COVER CUP
SMEARED CUP

4. Remove the cover cup and expose the jellied cup in places where you might trap smog particulates. Label the cups with your name and smog location.

5. Take care where you place the cups. They should be exposed for two to five days. Use the covering cup to protect them when they are not in a smog location.

Where do you plan to expose your cups?

1. _____   2. _____

3. _____   4. _____

Organize Your Smog Data

1. Observe your cups after two to five days. You will see many large and small particles. It is impossible to count them all. Ignore the very small ones and count only the reasonably sized particles. Our object in this investigation is not to get a total count but to compare your various cup locations. If you are in a low smog area, use a hand lens to make the count.

2. Fill out the Smog Particulate Data Table below. Let two friends help you count the particles on each cup. Record the counts.

3. Average the three counts and record the average for each of the cups. Round off the averages to the nearest whole numbers.

SMOG PARTICULATE DATA TABLE				
LOCATION OF CUP	**PARTICLE COUNT AFTER 2 TO 5 DAYS**			
	YOUR COUNT	**FRIEND'S COUNT**	**FRIEND'S COUNT**	**AVERAGE** (ROUNDED OFF)

Name _____

Which location had the most particulates? _____

_____

Which location had the least particulates? _____

_____

4. Draw a few of the most interestingly shaped particulates. A hand lens would help.

Newton Smog Extra

Windshield Smog

1. Wash the glass windshield of your family car super clean.

2. Have your parents leave the car outside for a full day.

3. Obtain a clean, white rag and wipe the windshield vigorously.

Describe the appearance of the white rag. _____

_____

Unshiny Metal

Caution! Only do this with the help of a teacher or adult!

1. Obtain a piece of shiny metal.

2. Hold it at arms length with a tool and expose it to the exhaust of a running car for a **few** minutes.

Caution! Turn your face away and avoid the auto fumes.

Describe any changes in the shiny metal. _____

_____

What does this experiment tell you about autos and smog?_____

_____

Bully for Bernoulli

Newton Wants You to Know

Over 200 years ago, a Swiss scientist named Bernoulli experimented with moving air. He discovered that moving air exerts less pressure than still air. Bernoulli, of course, had never heard or dreamed of an airplane. Yet, his moving air ideas would eventually be used to build the wings that lift airplanes off the ground.

Here is Bernoulli's principle stated simply. When air speeds **up**, its pressure goes **down**. This principle can be used to explain all of the experiments in this activity.

Bernoulli Fun

A Paper Wing

1. Cut a sheet of notebook paper in half the long way.

2. Hold the paper at one end just below your mouth.

3. Blow over the top of the paper.

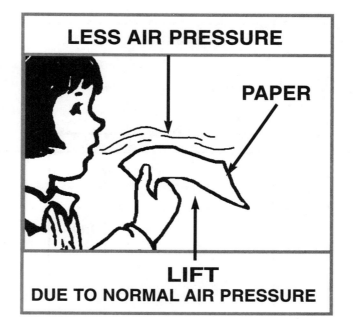

LESS AIR PRESSURE

PAPER

LIFT
DUE TO NORMAL AIR PRESSURE

Name _____

Can you explain what happened in terms of Bernoulli's principle? _____

_____

_____

Which Way Will It Move?

1. Attach strings to two small Ping-Pong™ or Styrofoam™ balls.

2. Mount them on a ruler so that they hang evenly.

3. Adjust them so that they are about 1/2" (1.25 cm) apart.

4. Use a straw to blow a stream of air between the two balls. What happens? Can you explain the result in terms of Bernoulli? _____

_____

_____

_____

_____

RULER

ABOUT 8" (20 cm) OF STRING

BALLS ARE ABOUT 1/2 INCH APART

STRAW

BLOW HERE

I Won't Budge

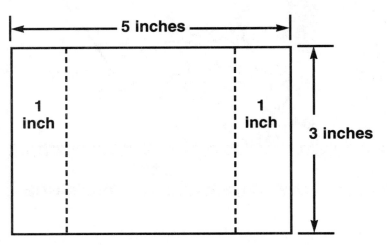

5 inches

1 inch

1 inch

3 inches

1. Obtain a 3" x 5" (8 x 13 cm) file card, or cut a piece of thin cardboard that size.

2. Mark off lines 1" (2.5 cm) from each end.

Name _____

3. Fold both sides at right angles along the 1" (2.5 cm) mark. You now have a small paper table.

4. Place your cardboard table on a real table as shown.

5. Blow **underneath** your cardboard table as hard as you can.

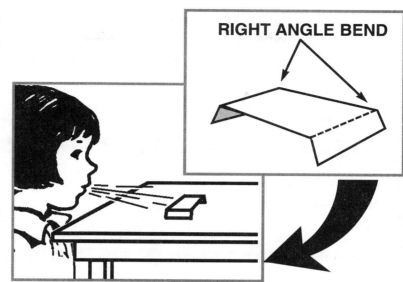

RIGHT ANGLE BEND

What happened? Can you explain this by Bernoulli's principle? _____

_____

_____

Floating on Air

1. Obtain a light Styrofoam™ ball and a hair dryer.

2. Point the dryer nozzle **exactly** straight up.

3. Turn the air on low or high.

4. Carefully and slowly place the ball about 6" (15 cm) above the blast of air.

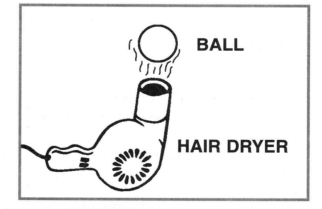

BALL

HAIR DRYER

Caution! Be careful with the hot air!

Release the ball and explain what happens. _____

_____

_____

Name _____

5. Keeping the dryer upright, slowly move it to the left or right.

What happened to the ball as you moved the dryer? _____

_____

Newton Hint: Observe the floating ball carefully. It never stays still but keeps moving from one side of the air stream to the other. As it tries to move left out of the airstream, the right side has **faster air**. Faster air means less air pressure. This allows normal air on the left to push the ball back into the airstream. Every time the ball tries to escape, Bernoulli's principle pushes it back in.

Newton Explains Airplane Wings

You are now experts on Bernoulli's Law. You know that when air speeds up, its pressure goes down.

Observe the model airplane wing on the right. It is curved on the top and has a flat bottom. Air speeds up as it passes over the curved surface so pressure on top of the wing goes **down**. Air speeds up much less on the flat surface. Therefore, it has more pressure than the wing top. This results in a force called **lift** that keeps airplanes up.

Real airplane wings have curves on both top and bottom to cut down on turbulence. The top of wings, however, are curved the most.

PATH OF AIR ABOVE WING
distance and speed greater; pressure less

CROSS-SECTION of AIRPLANE WING

PATH OF AIR UNDER WING
distance and speed less; pressure greater

Name _____

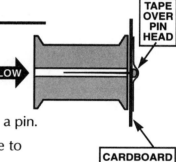

NEWTON'S
ACTION LAB
Air Science
15

More Fun with Bernoulli

Newton Reviews Bernoulli

Bernoulli's Law states that when air speeds up, its pressure goes down. Airplanes get their lift from the air speeding up across curved wings. Baseball pitchers use Bernoulli's Law to give their spinning balls a curve. Paint sprayers and automobile carburetors also obey Bernoulli's Law.

Bernoulli Fights Gravity

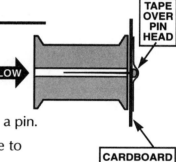

TAPE OVER PIN HEAD

BLOW

CARDBOARD

1. Obtain a spool with one hole, a 3" x 3" (8 x 8 cm) piece of light cardboard and a pin.
2. Place the pin in the center of the card. Secure the head of the pin with tape to hold it in place.
3. Place the pin in the hole of the spool as shown.
4. Hold the spool in front of your mouth and blow air through it in a steady stream.

Describe what happened. _____

Why didn't the card blow away? **Newton Hint:** The air between the spool and cardboard sped up. _____

Try the experiment again using this variation.

5. While blowing through the spool, turn your face down.

Why did the card defy gravity as you blew downward? _____

Bernoulli and a Candle Flame

Cars and trains are **streamlined** to make the air flow by them with the least resistance. They are given a teardrop shape to accomplish this. Can you imagine riding in a car with a flat front?

TLC10142 Copyright © Teaching & Learning Company, Carthage, IL 62321-0010

Name _____

Bernoulli plays a part in streamlining. Here is a way to demonstrate this for yourself.

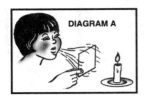

1. Obtain a candle and mount it **firmly** in a sturdy base.

2. Obtain a piece of cardboard and a soda pop can that is unopened.

3. Have an adult light the candle and supervise.

4. Set up the experiment as in diagram A and blow at the center of the card.

   Which way did the candle flame move? _____

   _____

   Why did the flame blow in the direction it did? **Newton Hint:** Your fast moving breath flowing around the card edges created a low pressure area on the candle side of the card. _____

   _____

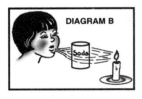

5. Repeat the experiment using a soda pop can as shown in diagram B.

   Which way did the flame move? _____

   Why did the flame move the way it did? **Newton Hint:** What did the streamlined shape allow your breath air stream to do? _____

   _____

Newton Wants You to Stretch Your Mind

Bernoulli's principle even explains how pitchers throw curves. You can throw your own curves using a mailing tube (or rolled up cardboard) and a Ping-Pong™ ball.

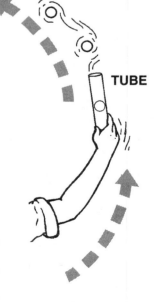

TUBE

1. Seal one end of a small mailing tube.

2. Place a Ping-Pong™ ball inside.

3. Extend both your arm and the tube.

4. Pitch the ball out of the tube as hard as you can. Practice will help you throw better curves.

Why does the ball curve? You are on your own to explain this Bernoulli experiment. _____

_____

_____

Newton Hint: The ball would not curve if it wasn't spinning.

NEWTON'S
ACTION LAB
Air
Science
16

Flying Through Air

Newton Explains How Airplanes Fly

Airplanes fly by obeying the laws of **aerodynamics**. Aerodynamics is a branch of physics that studies the flow of air around and against objects. Airplane designers must follow the laws of aerodynamics or their planes won't get off the ground.

There are four forces involved in flying an airplane. They are shown at the right. Gravity pulls the airplane down while lift works against gravity to push it up. Thrust pushes the plane forward, while drag tends to slow it down.

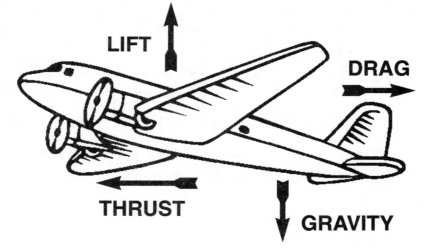

LIFT

DRAG

THRUST

GRAVITY

If a plane is to fly, the positive forces of lift and thrust must exceed the negative forces of gravity and drag. The engines provide the thrust while the shape of the wings provide the lift.

Flying Ring Competition

The aerodynamic background above will help you build a better **flying ring**. A flying ring is a simple device shown on page 40. Can you build one that can fly faster and straighter than your friend's? Enter the flying ring contest to be an aerodynamic winner. Please note that the ring dimensions are in metric centimeters.

Name _____

Flying Ring Contest Rules

1. All parts of the flying rings must be made of paper products. Thin cardboard, construction paper and straws work fine. No heavy weights can be used.

2. All flying rings must be built and tested at home. Only five minutes will be allowed during the contest for repair and for practice flights.

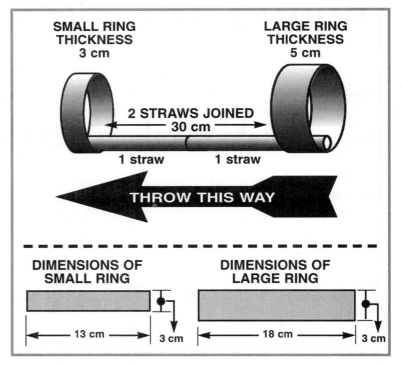

3. Flying rings will be judged on distance flown and ability to fly straight.

4. The dimensions must be as close as possible to the dimensions shown. All flying rings will be checked before flight for correct dimensions.

5. You may join others to form a flying ring team.

6. Mark your flying ring with a symbol or name for identification.

7. The Flying Ring Contest will take place on _____.

Newton Wants You to Build a Better Flying Ring

Newton hopes you enjoyed his standard flying ring. Now he would like you to design and build your own flying ring. You may make any variation you want from the basic model. You can change length, size of rings, number of rings. Anything goes.

Good Luck!

Human Breathing

Newton Wants You to Know

You can breathe through your mouth or nose. It is best to breathe through your nose. The nose has tiny hairs that filter out dust in the air. Your nose also warms and moistens the air before it gets to the lungs.

Air passes through the windpipe (trachea) on the way to the lungs. The windpipe divides into branches that serve the two lungs. Your vocal cords are at the top of the windpipe. Air must pass through the vocal cords to make sound.

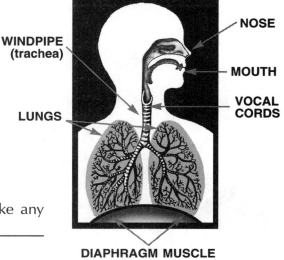

1. Close your mouth and squeeze your nose. Try to make any sound including humming. What were the results? _____

_____

_____

Adult lungs are about the size of footballs. They are hollow inside and have a surface area 30 times greater than your skin. Lungs exchange the oxygen your body needs with the carbon dioxide waste your body must get rid of. Blood comes to the lungs to exchange carbon dioxide for oxygen.

Twelve pairs of ribs surround and protect the soft, spongy lungs. This rib cage can expand or contract as you breathe.

2. Hold your rib cage and breathe deeply. You can feel your rib cage expand.

AIR SYSTEM FACTS

- Each adult lung weighs about 1 pound (.45 kg).
- A normal person takes in about 3000 gallons (11,340 liters) of air each day.
- Air going into your lungs contains 21% oxygen and practically no carbon dioxide.
- Air leaving your lungs contains about 16% oxygen and over 4% carbon dioxide.

Name _____

Measuring Your Lungs

An adult man has room for about six quarts (5.7 l) of air in his lungs. An adult woman has room for five quarts (4.75 l). In normal breathing, you can only force about four quarts (3.8 l) of air in and out of your lungs.

Chest Expansion

1. Obtain a tape measure or a long string.

2. Breathe as deeply as possible. Measure the size of your chest. _____ inches (cm)

3. Exhale as completely as possible. Measure the size of your chest. _____ inches (cm)

4. How much larger was your chest when expanded? _____ inches (cm)

Measuring Your Lung Capacity

How much air can you force out of your lungs? Here is a way to find out.

1. Obtain a half- or one-gallon (1.88 or 3.78 l) clear plastic bottle, a large bowl and about 2' (.61 m) of rubber or plastic tubing.

2. Use rubber bands to attach a ruler to the bottle as shown.

3. Fill the bowl half full with water.

4. Fill the bottle completely with water.

5. Place your hand over the mouth of the bottle. Turn the bottle upside down in the bowl below the water level. Remove your hand. The water will stay in the bottle.

6. With the help of a friend, tilt the bottle and push the plastic tubing a few inches (centimeters) inside.

7. Keep the bottle tilted so that the tube is not pinched.

8. Inhale as much air as you can, and use the tube to blow as much air as you can into the bottle. Use only one breath.

9. Measure the amount of inches (centimeters) of water your breath displaced. _____ inches (cm)

10. Wash off the tube and repeat the breath experiment with friends.

Here is how you can get a rough idea of the quarts (liters) of air you expelled. Compare the inches (centimeters) of water removed to the total inches (centimeters) of the bottle. If you blew out half of the water in a one-gallon (3.78 l) bottle, your air volume was half a gallon or two quarts (1.9 l).

Name _____

How Your Diaphragm Muscles Work

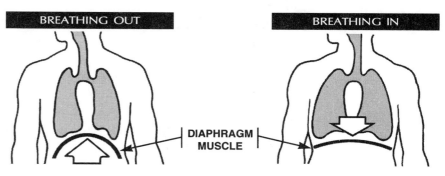

BREATHING OUT · BREATHING IN · DIAPHRAGM MUSCLE

Lungs cannot expand and contract by themselves to move air in and out. The powerful diaphragm muscle just below the lungs causes them to move. Pulling the diaphragm muscle down fills your lungs with air. Moving the diaphragm up forces air out of your lungs.

1. Breathe in and out deeply. You can feel the diaphragm muscle move.

2. Study the diagram above. It shows how diaphragm muscles move to help you breathe.

Building a Model Lung

You have learned that the diaphragm muscle helps you breathe. A lung model is shown on the right. Try to collect the materials needed to build it. You will need a one-hole rubber stopper, two rubber bands, plastic tubing, one large balloon, one small balloon and one common two-liter (¹/₂ gallon) plastic bottle.

glass or plastic tube (windpipe)
1-hole rubber stopper
rubber band
clear plastic bottle (chest)
balloon (single lung)
large balloon cut to fit (diaphragm)
strong rubber band

Note: Your real body has two lungs. This model is simplified by showing only one lung.

1. Be **careful** in cutting the bottom off the plastic bottle.

2. Move the rubber diaphragm in or out to affect the size of the "lung" (balloon).

WHY PEOPLE HICCUP, YAWN, COUGH OR SNEEZE

HICCUPS: When you eat or drink too much, your stomach stretches and irritates the diaphragm. This causes the diaphragm to contract swiftly.

YAWNS: When you are at rest, you breathe slowly. The lack of carbon dioxide triggers the brain to require an extra deep breath called a yawn.

COUGHS or SNEEZES: Dirt irritates the windpipe or nose. Coughs and sneezes exhale large amounts of air suddenly to remove the irritating dirt.

Name _____

Your Breathing Rate

Newton Explains Your Respiratory System

You can choose the food and water you drink. You have no choice about the air you breathe. Clean or dirty, air enters your lungs.

Dirty air can affect your health in many ways. It can irritate your eyes, nose and throat. It can injure your respiratory system and can shorten your life.

Your lungs branch out like a tree. At the very end of the branches are millions of air sacs called **alveoli**. Blood surrounds the outside of the alveoli to pick up oxygen and get rid of carbon dioxide gas.

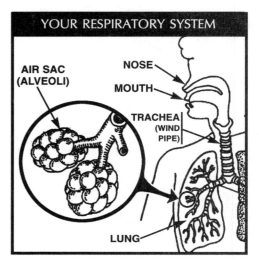

YOUR RESPIRATORY SYSTEM

AIR SAC (ALVEOLI)
NOSE
MOUTH
TRACHEA (WIND PIPE)
LUNG

Measuring Your Breathing

Your brain controls how fast you breathe. The carbon dioxide level in your blood triggers the brain. When you exercise vigorously, you produce more carbon dioxide. This signals the brain to have you breathe faster.

1. Sit still for two minutes.

2. While still sitting, count your breathing for one minute. Count inhaling and exhaling as one breath. _____ breaths

3. Now walk in place or walk around the room for two minutes.

4. Count your breathing rate for one minute. _____ breaths

Name _____

5. Now run in place or around the room for two minutes.

6. Count your breathing rate for one minute. _____ breaths

7. Try these breathing rate experiments on some friends. How do their rates compare to yours? _____

_____

_____

Most people inhale about one quart (.95 liters) of air with every two breaths. Let's use this fact to do some breathing math.

8. Divide your at-rest breathing rate by two. This gives you the quarts (liters) of air you breathe each minute. _____ quarts (liters) per minute.

9. Multiply the quarts per minute by 60. This gives you the quarts (liters) of air you breathe each hour. _____ quarts (liters) per hour.

10. Multiply the quarts per hour by 24 to get the quarts (liters) of air you breathe each day _____ quarts (liters) per day.

Suppose you had to pay five cents for each quart (liters) of air you need in one day. How much would air cost you per day? $_____

Bad Breath Special

The air you breathe in is about 21% oxygen. The air you breathe out is only about 16% oxygen. Your body uses the oxygen to burn the food it needs. When your cells burn food, they produce carbon dioxide (CO_2).

The air you breathe in has less than $1/10$ of 1% CO_2.

The air you breathe out is almost $4^1/2$ % CO_2. The CO_2 you exhale is only a small part of the ingredients of polluted air.

What effect can breathing polluted air have on your breathing rate? You can find out by breathing into a small paper bag. The high level of CO_2 in the bag will take the place of smog-filled air.

Caution! Students with asthma or other breathing conditions may need to skip this experiment. Skip this experiment if it makes you feel bad.

Name _____

1. Place a small **paper** bag tightly over your mouth and nose as shown.

2. Breathe into the bag for at least one full minute. This will increase the CO_2 you inhale.

3. Remove the bag and count your breaths in one minute. _____ breathing rate with CO_2.

Compare this rate with your previous normal rate. What happens to your breathing rate when the amount of CO_2

increases? _____

_____

_____

Newton Wants You to Explore Further

COMPARING BREATHING RATES	
ANIMAL	**BREATHS PER MINUTE**
Newborn child	55
Teenager	20
Adults	16
Giraffe	32
Horse	10
Rat	86
Rabbit	37

Your Pets Also Breathe

Do you have a pet?

Try to count and compare its rest and exercise breathing rate.

Smog and Your Respiratory System

The following substances are found in smog. Find out how they get into the air and how they can affect your respiratory system.

carbon monoxide   sulfur dioxide

nitrous oxides   ozone

hydrocarbons   pollen

Name _____

NEWTON'S
ACTION LAB
Air
Science
19

Air Helps You Hear and Talk

Newton Explains Your Ear and Air

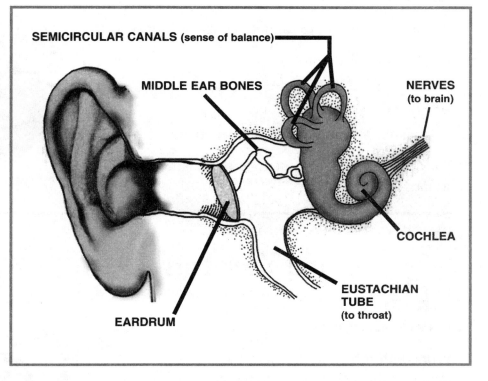

There would be no sound without air. All sound is due to vibration. When you talk, your vocal cords vibrate the air. Air vibrations from sounds move into your ear and vibrate your eardrum. This vibrates three tiny bones called the **hammer**, **anvil** and **stirrup** in your middle ear.

SEMICIRCULAR CANALS (sense of balance)

MIDDLE EAR BONES

NERVES
(to brain)

COCHLEA

EARDRUM

EUSTACHIAN
TUBE
(to throat)

The vibrating bones cause vibrations inside the **cochlea**. This is a snail-like structure filled with fluids and nerve endings. The nerve endings connect to the brain which interprets the sounds you hear.

The **semicircular canals** above your inner ear are responsible for your sense of balance. The **Eustachian tube** connects the middle ear to your throat. It serves to equalize air pressure inside and outside your ear. You may have felt the pain of unequal pressure in your ear while flying in an airplane.

47

Name _____

Experimenting with Your Vocal Cords

Feeling Your Vocal Cords

1. Feel the lump in your throat that is called the Adam's apple. It is your voice box and contains the two thin muscles of your vocal cords.

2. Hum loudly as you touch your vocal cords. You will feel them vibrating to make sound.

Balloon Vocal Cords

1. Blow up a balloon to about the size of your head.

2. Hold each side of the balloon with your thumb and forefingers as shown.

3. Stretch the mouth of the balloon as the air escapes to make high and low sounds.

Your vocal cords also stretch to make different sounds.

Testing Your Ears

How Far Can You Hear?

Work with others to measure the distance at which you can no longer hear a standard sound. The standard sound could be any of the below:

a. a ticking watch

b. a rubber band around a box

c. a clicking ballpoint pen

d. a marble dropped into a can

e. a radio set at a low level

f. a sound maker of your own invention

48

Name _____

1. Measure and record the distance at which people stop hearing the standard sound.

   Person 1 _____ ft. (m) Person 2 _____ ft. (m)

   Person 3 _____ ft. (m) Person 4 _____ ft. (m)

2. Check your friends, parents and grandparents to find if age affects hearing distance ability.

It Takes Two Ears to Determine Sound Direction.

Use your standard sound to check your ability to identify the direction of sound.

1. Close your eyes. Have a friend make a sound from various locations around you. Can you point to the sound's location?

2. Close your eyes again.

3. **Cover one ear** with your hand so that no sound can enter.

4. Have a friend again make sounds from various directions. Can you detect the location of the sound better with one or two ears?

Newton's Favorite Sound Ideas

Fool Your Friends

Collect or record some different kinds of sounds. Can your friends identify them?

Research Eardrum Pain

You learned on page 47 why your ears hurt in an airplane, on a mountaintop or even going up in a fast elevator. Find out what you can do to minimize eardrum pain.

EAR FACTS

* The pitch of a sound is measured in cycles per second.
* You can hear sounds between 15 and 20,000 cycles per second.
* Dogs can hear sounds that you can't. They can hear sounds as high as 50,000 cycles per second.
* Bats and some moths hold the hearing record. They can hear sounds up to 100,000 cycles per second.

How Animals Breathe

Newton Explains Why Animals Need Plants

OXYGEN CARBON DIOXIDE CYCLE

CO_2

O

Plants are useful to us for more than food, clothing and beauty. Animals take in **oxygen** from the air and give off **carbon dioxide**. Plants, in turn, take in our waste gas (carbon dioxide) and return oxygen to the air. This is called the **oxygen-carbon dioxide cycle**.

Every green leaf and every blade of grass is part of this essential cycle. Without this cycle, no animal could survive.

How about showing more respect for green living things? They make your life possible. Don't step on that grass.

Testing Your "Bad" Breath

So far you have taken Newton's word that you pollute the air with carbon dioxide. Scientists test for CO_2 with a fluid called bromthymol blue. It turns green or yellow when CO_2 is blown through it. Let's use bromthymol blue to test your breath.

Bromthymol blue can be obtained from college or school labs, science supply stores and through science catalogs.

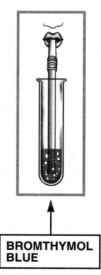

BROMTHYMOL BLUE

1. Fill a test tube $1/4$ full of bromthymol blue

2. Blow **gently** and **slowly** into the fluid. Don't blow so hard that the fluid overflows.

3. What color did the bromthymol blue turn? _____

What does this prove is in the air you exhale? _____

_____

Name _____

How Animals Breathe

GRASSHOPPER

Air enters pores

EARTHWORM

Oxygen passes through moist skin into blood

Oxygen passes through moist skin

Air passes into lungs

FROG

You breathe through your mouth and nose. Many animals do the same. Some animals breathe in very unique ways.

Grasshoppers, and most insects, breathe through openings (pores) in their body wall. Earthworms can take in oxygen through their skin. Frogs can take in oxygen through both their mouth and their moist skin.

Fish live in water. They get the oxygen they need from the small amount of oxygen normally dissolved in water.

Fish breathe by gulping water into their mouth and out their **gills**. Gills are fleshy areas covered by a bone flap. Oxygen passes through the gills into the blood system.

Let's try to observe and measure fish breathing.

1. Obtain a **small** goldfish and place it in a big jar of water.

2. Observe how the mouth opens and the gills move on each breath.

FISH

Water passes over gills

Oxygen goes into the blood

Water enters through mouth

3. How many times did your goldfish "breathe" in one minute? You may want to try this a few times and obtain an average. _____ breaths per minute

Name _____

Goldfish at Work

1. Obtain two **small** baby food jars with lids.

2. Obtain one **small** goldfish.

3. Obtain some bromthymol blue. Remember that this fluid turns green or yellow when CO_2 is added.

4. Fill one jar with water and enough bromthymol blue to make a very blue color.

5. Pour half of the fluid into your second jar. At this point both jars should have roughly the same amount of fluid.

6. Place the small goldfish in one jar.

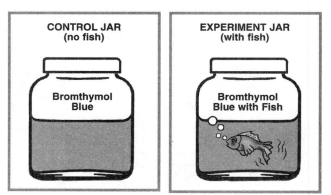

CONTROL JAR
(no fish)

Bromthymol
Blue

EXPERIMENT JAR
(with fish)

Bromthymol
Blue with Fish

Don't worry about the fish. It will be out of the small jar in 10 to 15 minutes.

7. Cover both jars with their lids.

8. Wait and observe your small goldfish for 10 to 15 minutes. If it doesn't look well, remove the poor fish to a bigger container. The jar without the fish is your "control." Scientists use controls to make their experiments more reliable.

What happened to the color in the control jar? _____

_____

What happened to the color in the fish experiment jar? _____

_____

What gas does this experiment prove is being given off by your goldfish? _____

_____

How Plants Breathe

Newton Knows Plants Are Alive

Plants are very much alive. They may not appear to move, but they do move as they grow. Plants are able to grow and reproduce. Plants need energy, take in nutrients (food) and breathe constantly.

Plants breathe through pores in their leaves called **stomates**. Observe the stomates shown. They can open or close as needed by the plant. Stomates also serve to release water from plants into the air. It is at the stomates that plants take in carbon dioxide and give off oxygen.

You can observe stomates for yourself if you have a microscope available. Here are some helpful hints in observing plant stomates.

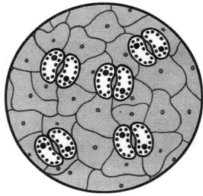

1. Select fleshy leaves like geraniums.

2. View the underside of the leaves.

3. Use tweezers to pull off the **thinnest** section of the leaf that you can.

4. Mount the thin leaf section on a glass slide in a drop of water.

Happy stomate hunting!

STOMATE FACTS
• Leaves may have up to 100,000 stomates per square inch (centimeters). • Stomate openings for air are only $1/20$ as thick as a sheet of paper. • Each stomate has two guard cells that control its opening.

Plants Give Off Oxygen

1. Obtain an aquarium plant called **elodea**. Any similar water plant will do.

2. Cut off a small portion of the stem.

Name _____

3. Place it with the **cut stem up** in a jar or glass of water. The water level should be at least ¹/₂" (1.25 cm) inch above the cut stem.

4. Place the jar outside or inside near a bright window or light.

5. Wait a short time and then observe the cut end of the stem.

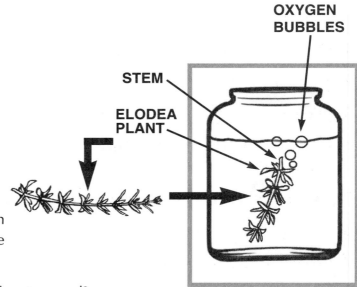

What do you see forming at the stem end?

_____

_____

What gas is made by green plants that animals need? _____

_____

Scientists, like Newton, have collected the gas given off by plants. When enough is obtained, they burn it to prove that the gas is oxygen.

Newton Loves Bromthymol Blue

You have already used bromthymol blue and elodea plants. Newton wants you to combine them into a controlled experiment. You are going to prove that plants take in carbon dioxide.

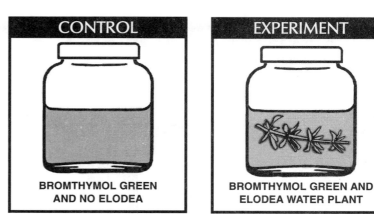

Name _____

To prove that plants take in carbon dioxide:

1. Obtain two similar jars, elodea and bromthymol blue.

2. Fill one jar **completely** with water.

3. Add enough bromthymol blue to make the water blue.

4. Use a straw to blow slowly through the fluid. It will turn yellow or green as your breath adds carbon dioxide.

5. Pour half of the fluid into the other jar which will act as your control.

6. Add lots of elodea to your second jar only.

7. Cover both jars and place in bright light.

8. Observe both jars after five to 10 minutes.

Did the color of the control jar change? _____

What color did the experiment jar change to? _____

What gas did you breathe into the bromthymol blue to turn it green? _____

What gas must the plant have removed to make your fluid blue again? _____

 # Newton Wants You to Think

Someday space travelers may be going on long voyages to distant planets. Can you give four good reasons to take plants along on their trips?

1. _____

_____

2. _____

_____

3. _____

_____

4. _____

_____

Name _____

Carbon Dioxide: An Essential Part of Our Air

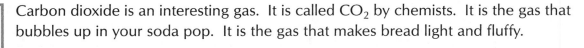

Newton's Famous Bread Recipe

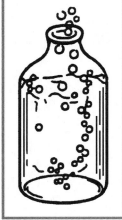

Carbon dioxide is an interesting gas. It is called CO_2 by chemists. It is the gas that bubbles up in your soda pop. It is the gas that makes bread light and fluffy.

Newton's favorite bread recipe requires carbon dioxide as an ingredient. Some bread makers use yeast combined with sugar to make CO_2. Newton uses baking powder to supply CO_2. Try his recipe.

1. Place three tablespoons (45 ml) of flour in a container.

2. Add one teaspoon (5 ml) of baking powder.

3. Add just enough water to make it doughy.

4. Place it in a warm oven and heat **gently** till the dough rises.

Describe how your "bread" looked and felt. _____

_____

What gas caused the empty spaces within your bread? _____

Newton Lectures on Carbon Dioxide

Carbon dioxide makes up only 0.03% of the air around us. Scientists can compress CO_2 into a solid called **dry ice** that is used for keeping things cold. CO_2 is not like water (H_2O). Water goes from a solid to a liquid to a gas. CO_2 is rarely found as a liquid. Dry ice goes directly from a solid to a gas form without becoming a liquid. This process is called **sublimation**.

Name _____

Newton Warning: You have learned that CO_2 is used in soda pop and bread and as dry ice for cooling. Don't mistake relatively harmless CO_2 for a gas called **carbon monoxide**. It has the symbol CO. Carbon monoxide can kill you. It comes from auto exhausts and poorly ventilated gas heaters.

The Carbon Dioxide Oxygen Cycle

All animals take in oxygen and give off carbon dioxide. All green plants take in carbon dioxide and give off oxygen. This cycle makes life on Earth possible.

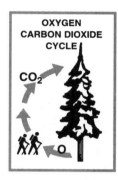

OXYGEN CARBON DIOXIDE CYCLE

CO_2

O

Animals Give Off Carbon Dioxide

1. Obtain a solution of bromthymol blue. This chemical changes color if CO_2 is added.

2. Pour enough into a test tube to make it ¹/₄ full.

3. Use a straw to breathe **gently** into the fluid. Remember that you are an animal.

What color did the fluid become? _____

_____

Save the changed bromthymol blue for the plant experiment below.

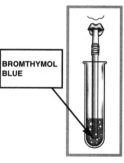

BROMTHYMOL BLUE

Plants Absorb Carbon Dioxide

1. Place the used bromthymol blue in a small jar. A baby food jar is perfect.

2. Add a few aquarium plants such as elodea to the jar.

3. Cover the jar and place it in sunlight or near any light source.

What color did the fluid become? _____

You used your breath as a source of CO_2 to change the bromthymol blue color. What must the plants have **removed** from the fluid to change it back to blue?

_____

BROMTHYMOL GREEN AND ELODEA WATER PLANT

Name _____

Mysterious Carbon Dioxide

Discovering the Nature of Carbon Dioxide

Carbon dioxide gas has **three** main characteristics that cannot be observed directly. Newton was highly skilled in observing the nature of things. What three things did Newton learn about carbon dioxide after he did the experiment below?

1. Obtain a tall candle. Place it in a **solid** base. Do not use a birthday candle.

2. Obtain baking soda, vinegar and a quart (liter) jar.

3. Obtain an empty, small (6 oz. [170.1 g]) fruit juice can.

4. Use a can opener to remove only the bottom end. **Be careful with sharp edges**.

5. **Carefully** use a pair of pliers to widen the opening at the drinking end. Make the opening slightly smaller than your candle's diameter. **Carefully** force the candle halfway into the can.

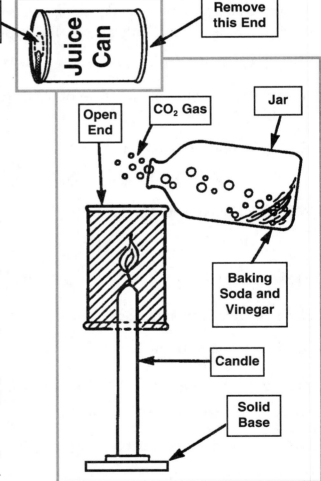

Widen Opening

Juice Can

Remove this End

Open End

CO_2 Gas

Jar

Baking Soda and Vinegar

Candle

Solid Base

Name _____

6. Have an adult light the candle with caution on top of a sink. **Think safety!** You may need to tilt the candle to light it.

7. Place two teaspoons (10 ml) of baking soda in the jar.

8. Add a **small** amount of vinegar to the jar. Do this on top of a sink. The gas bubbling up is CO_2.

9. Hold the mouth of the jar, as shown, over the top of the juice can. **Do not allow fluid** (vinegar) **to drip onto the flame.**

Describe what happened to the flame. _____

_____

_____

10. Newton learned **three** important things by observing CO_2 gas indirectly. Can you list the three things you and Newton should have learned from this experiment?

　　1. _____

　　2. _____

　　3. _____

Carbon Dioxide Thought Problem

You are locked in a small room with a huge piece of dry ice (solid carbon dioxide). It is warm and the dry ice is rapidly turning into a gas that can smother you.

Where is the best place in the room for you to wait for help? _____

_____

_____

_____

NEWTON'S
ACTION LAB
Air
Science
24

Siphon: An Air Pressure Device

How Does Newton Empty His Aquarium?

Newton uses a siphon to empty his aquarium. A siphon is simply a tube connecting water at a **high** level to water at a **lower** level.

To empty his aquarium, Newton fills the tubing full of water. He pinches both ends with his fingers. He places one end in the aquarium and releases his fingers. The other end empties into a jar **below** the aquarium.

This same idea can be used to siphon water from a reservoir over small hills to cities below.

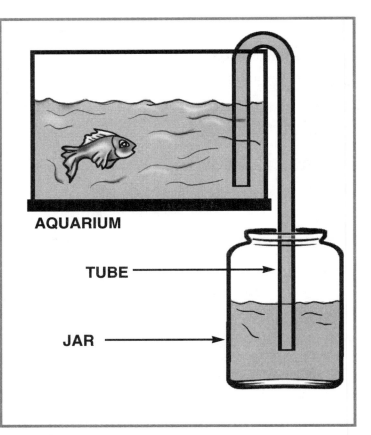

AQUARIUM

TUBE

JAR

Siphons work by **air pressure**. As the water flows through the tube to the lower jar, it reduces the pressure above it. Normal air pressure on the upper jar keeps the siphon working.

SIPHON LAWS

- Siphons always flow from higher to a lower level.
- The greater distance between levels, the faster the flow.

Name _____

Demonstrating Siphon Laws

1. Obtain two tall jars and some tubing.

2. Fill one jar with water.

3. Fill the tubing with water, pinch both ends and use Newton's technique to start the siphon.

> A siphon can also be started by sucking water through it.

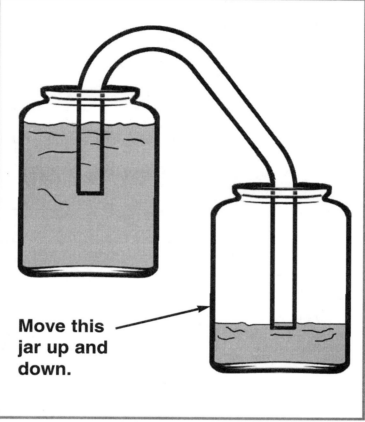

Move this jar up and down.

4. Let the water flow. Now try these variations:

   a. Place one jar as low as possible below the other.

   b. Place the jars barely below each other.

   c. Place the jars even with each other.

   d. Reverse the heights of the two jars to see if the siphon will go backward.

Refer back to the siphon laws on page 60. Explain what you learned about siphons

in terms of these laws. _____

_____

_____

_____

Name _____

Siphons Need Air Pressure

1. Obtain a small neck gallon (liter) jar, a two-hole stopper, a piece of glass tubing, a rubber tube and a smaller jar. Fill the large jar with water.

2. Set up the materials as shown.

3. Start the siphon by sucking on the tube and letting the water flow.

4. As it flows, cover the open stopper hole with your finger so that no air can enter.

What happens? _____

_____

_____

_____

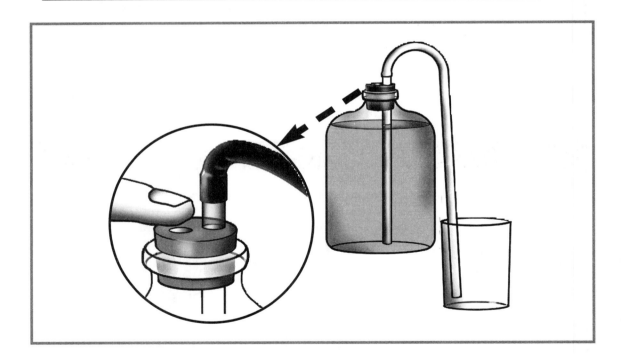

What does this experiment prove about siphons and air pressure? _____

_____

_____

_____

Name _____

Newton's Siphon Puzzlers

Newton likes to make you think. Here are two of his favorite siphon puzzlers. In both puzzles, put on your thinking cap to predict where the water will end up.

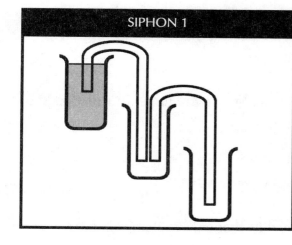

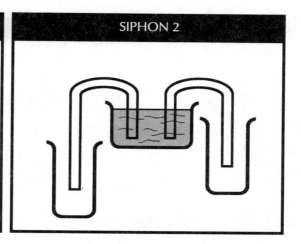

Sketch below what siphon 1 will look like when the water stops flowing.

Sketch below what siphon 2 will look like when the water stops flowing. Assume that the action starts at the same time in both tubes.

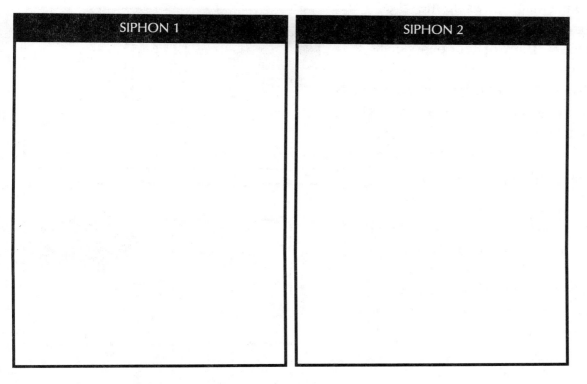

Answer Key

Air Gas Chart, page 8

1. nitrogen
2. oxygen
3. argon
4. carbon dioxide
5. neon
6. helium
7. krypton
8. xenon
9. hydrogen
10. water

I'm Falling for You, page 10

Plate B flips the least.

How Much Does Your Balloon Weigh? page 13

The weight of the air in the balloon is usually between 0.3 and 0.6 grams.

Balloon Puzzler, page 15

The smaller balloon actually blows up the larger balloon. This is because of the greater elasticity of the smaller balloon. Think of your experience blowing up balloons. At first, a balloon is difficult to blow up because of the rubber's initial elasticity. Then the balloon loses its elasticity and is easier to blow to a larger size.

Help! I'm Stuck in a Milk Bottle, page 22

The egg can be removed intact from the bottle by reserving air pressure. Place your mouth *completely* around the bottle. Tilt the bottle up so that the egg is resting on the bottle's mouth. Blow as hard as you can into the bottle. The added air pressure inside the bottle will force the egg out. If you are not speedy, the egg may land in your mouth.

Newton's Helpful Hint, page 24

Air pressure is working against you in this puzzler. It tends to keep the pop in the bottle. You must get air inside the bottle to help you force the pop out.

Hold the gallon (liter) bottle **firmly** with two hands over the sink. Rotate it rapidly to form a "tornado"-like air funnel in the bottle. Stop rotating and let the air come in and push the water out.

Discovering the Nature of Carbon Dioxide, page 59

10. CO_2 is invisible.

   CO_2 puts out fires.

   CO_2 is heavier than air.

Carbon Dioxide Thought Problem, page 59

Since CO_2 is heaver than air, it will collect near the floor. Your best breathing air is near the ceiling.

Newton's Siphon Puzzlers, page 63

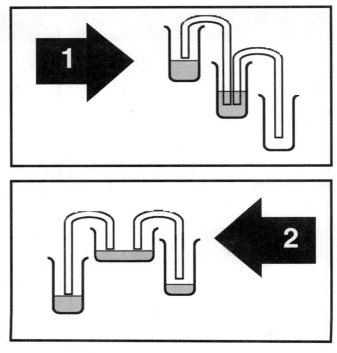